Listen • Read • Think

SCIENCE

Camouflage

Copyright © QEB Publishing, Inc. 2004

Published in the United States by
QEB Publishing, Inc.
23062 La Cadena Drive
Laguna Hills, CA 92653

This edition published by
Teacher Created Resources, Inc.
6421 Industry Way
Westminster, CA 92683

www.teachercreated.com

Library of Congress Catalog Card No. 2004101817

ISBN 978-1-4206-8146-8

Written by Terry Jennings
Designed by Zeta Jones
Editor Hannah Ray
Picture Researcher Joanne Beardwell

Series Consultant Anne Faundez
Creative Director Louise Morley
Editorial Manager Jean Coppendale

Printed and bound in China

Picture credits
Key: t = top, b = bottom, m = middle, c = center, l = left, r = right

Corbis/Linda Lewis 12–13, /Buddy Mays 4, /Roger Tidman 8; **Ecoscene**/Frank Blackburn
6–7, /Satyendra Tiwari 20–21, 23br; **Getty Images**/Daryl Balfour 14, /Wayne R Bilenduke
10–11, 22tl, /John Downer 15, /Nick Garbutt 18, /Peter Lilja 19, /Karen Su 16–17, 22bl, /Tom
Walker 9; **Still Pictures**/Yves Thonnerieux 5, 23tr.

Listen · Read · Think

SCIENCE

Camouflage

Terry Jennings

Teacher Created Resources

Some animals use their color to help them hide and keep safe.

This way of hiding is called camouflage.

The male pheasant is easy to see.

Can you see the female pheasant looking for food?

7

This animal is a weasel.

In summer, it is brown.

In winter, the weasel turns white. Why does it do that?

The polar bear lives in a place where it is always snowy.

The polar bear's white fur makes it hard to see her and her cubs against the snow.

The flounder has
a flat body.

Can you see
this fish when it
is on the bottom
of the sea?

13

Can you see the lioness
in the long grass?

This lioness is hunting. Why
does she have to run fast?

16

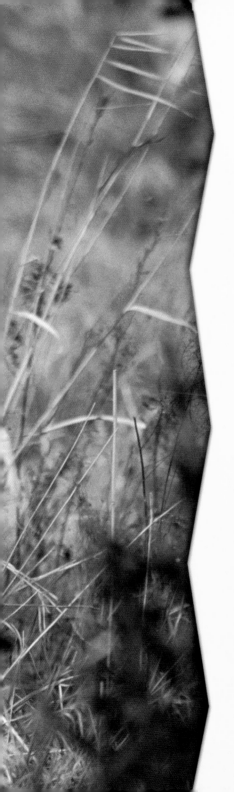

The tiger has stripes on its body.

The stripes help to hide the tiger in the long grass when it is hunting.

The chameleon can
change color quickly.

This means that it is always camouflaged.

20

The deer's brown
color makes it hard
to see against the
leaves in the woods.

Can you remember which animal is the same color as the snow?

Why does the tiger have stripes?

Can you spot the camouflaged butterfly? What does it look like?

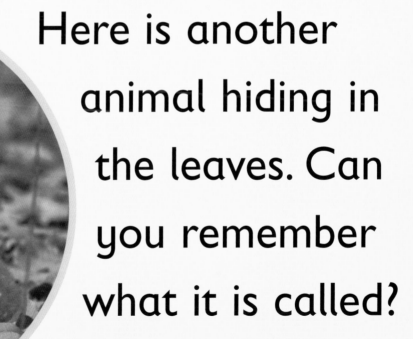

Here is another animal hiding in the leaves. Can you remember what it is called?

Parents' and teachers' notes

- Young children will need help pronouncing words such as "chameleon" and "pheasant," but once they have mastered these words, they enjoy repeating them. It may be necessary to explain the male and female pheasant in terms of "Daddy" and "Mommy".
- Look at pages 4 and 5 together. Can your child spot the camouflaged animals? (Stick insect and butterfly.)
- Look through the rest of the book. Does your child know the names of any of the animals? What color are the animals?
- What color are the animals that live in the snow? What color is the snow?
- Turn to pages 22 and 23. Look at each of the photographs and, together, talk about possible answers to the questions.
- Cut out paper butterflies in different colors and place them on a variety of colored surfaces around your home or classroom. Which are the easiest to see? Which are camouflaged?

- Paint some clean craft sticks green and others brown. Scatter some of the sticks on grass and some on dirt. Which are easier to see?
- Observe birds in your yard or at a park. Which of them have camouflage colors?
- Carefully turn over logs or large stones. Look for small animals that are camouflaged. Be sure to replace the logs or stones exactly as they were.
- Explore movement and ask your child to move like a polar bear, a fish, and a lion.
- Discuss why people such as birdwatchers need to wear clothes that camouflage them.
- Discuss why we shouldn't wear clothes that are camouflaged when we are crossing the road or walking along busy streets at night.